Seh and the Robot Friends

The Ingestible Origami Robot

18th March 2021

Dana Damian   Shuhei Miyashita   Ayzen Damian-Miyashita

and
For
annie

𝕭 ♡

x x Atzen

To Ayzen on his 5th birthday anniversary

It's a sunny day. Seh and her friends are playing outside.

After lunch Macrorobot is not feeling well. It has a tummy ache. It needs to take a rest.

Seh: 'I'll make a robot to repair your tummy. I'm cutting paper, shaping it like an origami, and folding it in the oven by heat.'

'We need a medicine inside the robot.
Mixing... Stirring...'

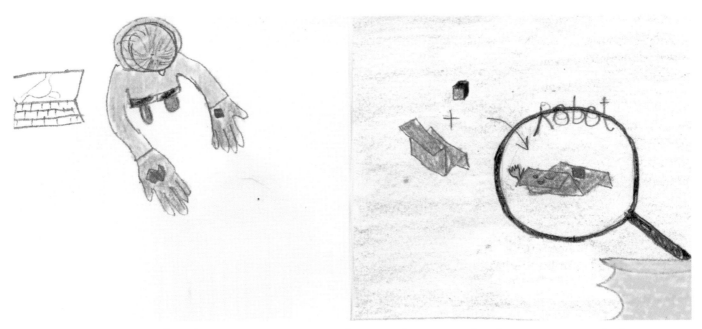

'And finally, a little magnet on the robot to help me track its path inside your tummy.'

Microrobot: 'Hm, let me see it closely through my magnifier. This robot is even smaller than me!'

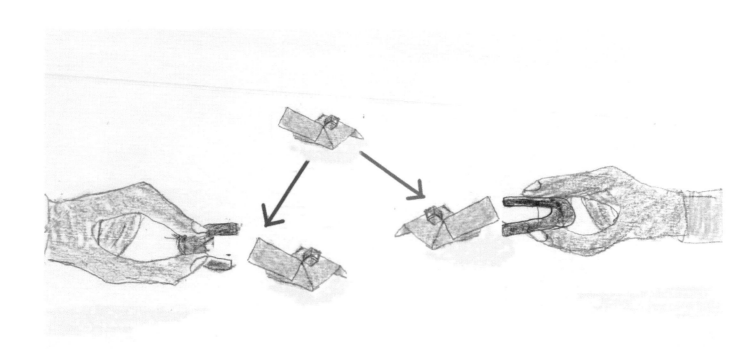

Seh: 'I am using magnets attraction to move the robot along a path. By hovering another magnet close to the robot, our robot will want to move toward that magnet.'

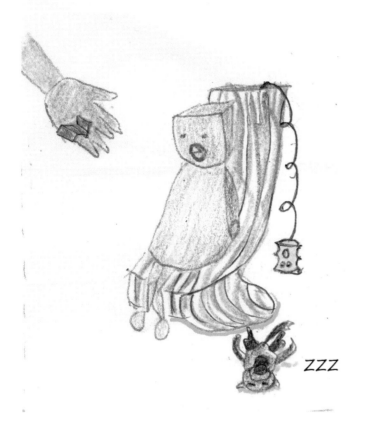

'The robot is ready!
You can swallow it now.'

ZZZ

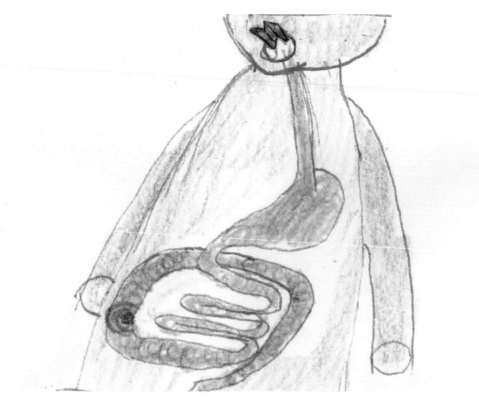

'The robot will travel through your tummy to find your wound and release the medicine.'

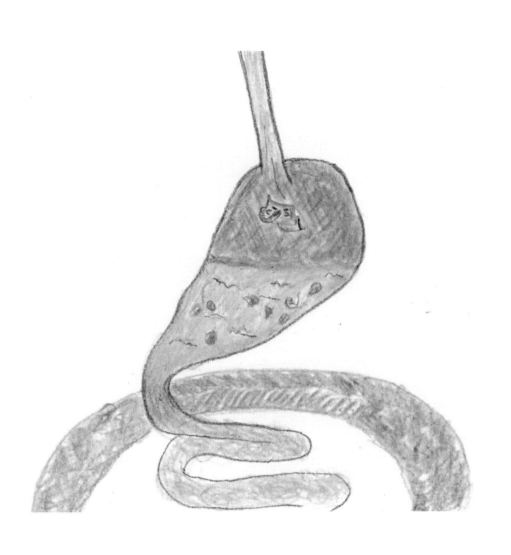

<< My mission is starting. And **3**, **2**, **1** : Jump in the pool! >>

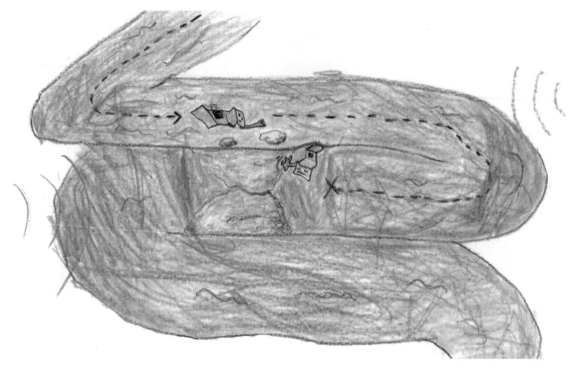

<< I have to smash this pile of food
blocking my way! >>

<< Let's dash to the target! >>

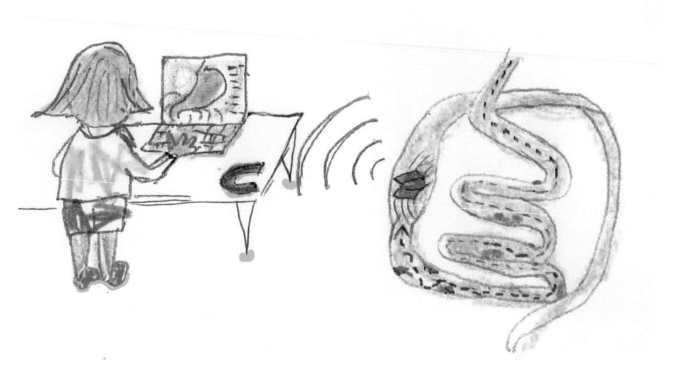

'The robot can look at and touch your tummy to feel
the wound, and then sends messages to my computer.'

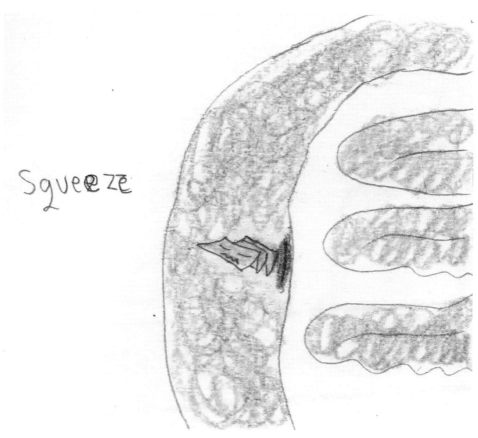

Squeeze

'The robot squeezes out the medicine onto the wound.'

Macrorobot: 'I feel much better! And the robot disappeared?!'
Seh: 'Yes, the robot biodegrades.'

Making robots is fun and useful.
The friends are up for more playing now.

The End

Definition: What is a robot?
A robot has:

3. Smartness
   when the sensors pick up
   changes in the surroundings
   the robot tells the
   motor how to move

Touch →

See →

1. Sensors
   to learn
   about the
   surroundings

2. Motors
   to move in
   or change the
   surroundings

4. Autonomy
   Robot charges
   (batteries) by
   itself, learns
   by itself

The ingestible origami robot has

1. Sensors
   touch soft or
   hard things
   see colors

2. Motors
   squeeze
   move through
   the tummy

3. Smartness
   avoid food
   in the tummy
   while moving;
   don't bump!
   in the wall

4. Autonomy
   learn about
   the tummy
   to move
   faster

Questions

1. What is Seh planning to do to help her friend Macrorobot?

2. Why origami is a good way to make the robot small?

3. What kind of challenges the origami robot finds inside the body?

4. What makes the origami robot move inside the body?

5. What materials may be good to make the ingestible origami robot?

Let's build a simple (non-ingestible) origami robot!

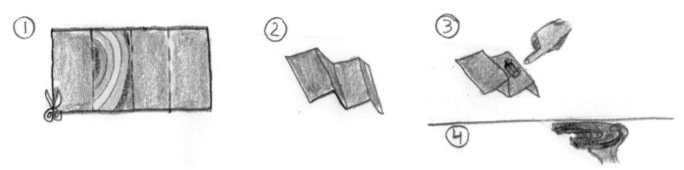

1. Draw a rectagle (see the red line) and cut it off. Draw the paper as you wish – a rainbow, a ☺. Then fold it along the dotted lines: the black dotted lines as mountain folds, and the blue dotted line as valley folds.
2. You will get an origami robot like in the story.
3. Add a little magnet on a top face of the folded origami. Place it on a table.
4. Take another magnet (a bigger one) and move this magnet under the table beneath the origami robot. Can you see the origami robot moving?
   You can play more with the folds and see if the origami robot changes the way it moves.

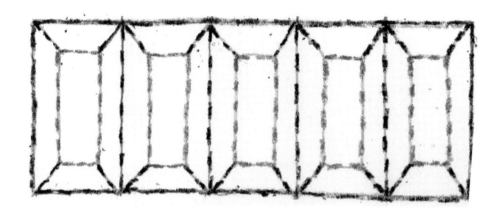

1. You can make a bit more complicated origami robot using this pattern. Draw the pattern on a piece of paper, cut the red rectangle. Then fold it along the dotted lines: the black dotted lines as mountain folds, and the blue dotted line as valley folds.

2. Attach the magnet on one of the blue rectangles. Move the origami robot using the second magnet. (Check if the origami robot moves differently if you change the position of the attached magnet.)

Draw on this page what kind of robot or machine you would like to build!

About the story
This story is adapted from:
Shuhei Miyashita, Steven Guitron, Kazuhiro Yoshida, Shuguang Li, Dana D. Damian, and Daniela Rus, **Ingestible, Controllable, and Degradable Origami Robot for Patching Stomach Wounds**, *IEEE International Conference on Robotics and Automation (ICRA)*, pp. 909-916, 2016

Written by Dana Damian and Shuhei Miyashita
Illustrated by Dana Damian and Ayzen Damian-Miyashita

Edition: 18th March 2021

Printed in Great Britain
by Amazon